IMAGES
of America

1906 SAN FRANCISCO
EARTHQUAKE

IMAGES
of America

1906 SAN FRANCISCO EARTHQUAKE

Richard Hansen and Gladys Hansen

ARCADIA
PUBLISHING

Published by Arcadia Publishing
Charleston, South Carolina

Printed in the United States of America

Library of Congress Control Number: 2012947056

For all general information, please contact Arcadia Publishing:
Telephone 843-853-2070
Fax 843-853-0044
E-mail sales@arcadiapublishing.com
For customer service and orders:
Toll-Free 1-888-313-2665

Visit us on the Internet at www.arcadiapublishing.com

To my wife, Karen, who puts up with my history interests.

CONTENTS

ACKNOWLEDGMENTS

We wish to thank the following people for their support and interest in the 1906 research: Diana Yee, for her support in helping to do research and handling email over 20 years; Dr. William Blaisdell, for opening our eyes to the medical problems of the disaster; and Kathleen Manning of Prints Old and Rare, who gave support and advice.

Lastly, we wish to thank all who have sent information on the 1906 disaster to our web site, the Museum of the City of San Francisco, at sfmuseum.org.

All images used in this book are from our own personal collection, which we have spent years cultivating.

INTRODUCTION

The early morning of April 18, 1906, was not any different from the previous morning, or the morning before that. Fishermen were coming in with the previous evening's catch, and bakers were beginning their labors, as they had for the days and weeks before. That morning, it took all of 85 seconds to destroy the gentle stillness associated with the wee hours of a normal San Francisco morning.

At 5:12 a.m., a 25-second foreshock shook the city awake. The foreshock, which would equal about a 4.0 on today's Richter scale, was only a gentle rumbling compared to what followed. Approximately 30 seconds later, the ground shook with a violence no one thought possible. San Francisco was hit by an earthquake of unbelievable intensity, registering about an 8.3 on the Richter scale. The quake lasted about 60 seconds and threw the thriving metropolis of the early 1900s into a chaotic mix of tumbling buildings and horrible fires, ending the life its citizens had come to know. The quake was felt from Oregon to Los Angeles and as far inland as Nevada.

The quake caused structural damage to most of the buildings and roads in the city. Many of the brick buildings simply collapsed into mounds, trapping whoever was unlucky enough to be inside. However, it was not the earthquake that caused the most damage. The quake ignited numerous fires throughout the city, which continued to burn for days until there was simply no fuel left.

It is estimated that more than 30 fires broke out, quickly overwhelming the understaffed fire department. The fires burned for 490 city blocks—approximately 25,000 buildings. Ironically, many of the fires were set by the fire department themselves. Using dynamite, they attempted to create "firebreaks" by leveling certain building or structures that were in the path of the oncoming blaze. Having not received proper training in the use of dynamite, this decision turned disastrous rather quickly, with many of the dynamited buildings catching fire and adding to the destruction. They raged for three days and nights.

In an instant, the great city of San Francisco was reduced to rubble, with flames scorching anything and anyone in their path. The total estimated damage from the quake and the subsequent fires came to almost $500 million. However, the emotional toll was incalculable. What started as a quiet morning turned into a nightmare, lasting long after the earth settled and the last flames were extinguished.

The following is a May 14 report from the fire alarm operator, James C. Kelly, to Chief W.R. Hewitt of the Department of Electricity in regards to the fire alarm system on April 18:

Dear Sir:
In reply to your request under date of the 11th Inst. for full detail as to matters happening under my observation the morning of the "Earthquake", I beg to say:
Shortly before sunrise, as I was standing at an east window of the Fire Alarm office [at 15 Brenham Place] looking toward the Hall of Justice, the earthquake began. I went at once to the clock to note the time and duration of the shock. The shock began at 5:13 plus 10 Secs. A. M., and I watched its duration at the clock for 19 seconds, when it became so severe, and there seemed such danger of the walls and ceilings of the building falling, that I ran to underneath the frame of one of the front windows as a place of greater safety;–the relay Oper. having meanwhile taken refuge in the doorway between the operating room and the battery room. The vibrations still continued for some seconds as I remained at

the window, and I saw meanwhile the southeast wall of the tower of the Hall of Justice fall; also part of the walls of a brick building near the corner of Washington Street and Brenham Place. Finally the shock ceased.

While yet the vibrations continued I had noted, by the running of the Registers on fire box signal circuits, that the lines on those circuits were open. Within a very few seconds after the shock ended I saw the smoke of an apparently large fire begin to rise from what I judged to be the vicinity of Market and Beale St.; (box 267). I at once went to key to strike out said box, (meanwhile calling to Relay Opr. to set up 267 on Repeater). No alarm came in for this fire, and be it noted that no alarms whatsoever came into the office after the commencement of the earthquake.

San Franciscans were skeptical about earthquake threats. The following essay, entitled, "Are we in Danger from Earthquakes?," appeared in the *San Francisco Real Estate Circular* in April 1872:

No disastrous earthquake has visited San Francisco within the past hundred years, and the series of earthquakes which lately occurred in the southeastern corner of the State, four hundred miles—on the edge of the great Interior Basin, and in a volcanic region—were by no means so disastrous or severe as those which visited the Mississippi Valley in 1811, or those which are of frequent occurrence in New Zealand, Japan and many other countries. Nevertheless, the people of the Atlantic States look upon California as one of the most dangerous earthquake countries in the world. This is due to the infernal habit of exaggeration which characterizes Eastern newspaper letter-writers and newspaper editors. The New York papers published such sensational heads as these, lately: "California Rocking from One End to the other!" "California's Fearful Cataclysm!" The Solid Earth Melting!" etc., etc., while a few correspondents here supplemented and added to the lies and excitement. The truth, however, is—judging by an analogy and all the light that science has placed within our reach—San Francisco is in very little more danger of a disastrous earthquake than the Eastern States of being flooded by an overflow of the Atlantic Ocean. Some of the distant southern and southeastern counties are subject to heavy shocks, but there is little danger, even there, to those who reside in frame buildings.

The Atlantic papers would most willingly publish any wild earthquake theory that San Francisco was in momentary danger of being swallowed up; but we have no hope that they will publish *these facts*. There would be no sensation in them, and nothing but sensation agrees with the feverish newspaper stomach there.

In the wake of the tragedy, the local press was full of stories of human drama. The following appeared in the *San Francisco Examiner* on April 29, 1906:

Wife he seeks may be dead, fireman saved baby but woman perished. James Fielding is searching for his wife and daughter in the little towns and rescue camps around the bay. A few days ago he arrived in the city from Arizona and set out on the quest for his family.

Yesterday Matthew Brown of Truck Company No. 8, reported to the Red Cross headquarters that he had charge of a little girl about 3 years old. Her mother had been killed in a burning building south of Market Street the day following the earthquake. An inspection of records showed that the little girl answered to the description of Fielding's missing daughter. The Red Cross nurses believe the woman who was killed was Fielding's wife. Brown will keep the child at his house in Oakland until Fielding calls.

Most everyone had stories to tell after the earthquake. It must have been difficult to abandon the treasures they loved enough to record so that they would never be forgotten. We are so glad they did so, even under great stress.

One

PRE-EARTHQUAKE AND FIRE

Early San Francisco seemed born to catastrophe, as disasters followed one after the other in the 1800s. Both fires and earthquakes were almost epidemic from the time of the city's founding to the ultimate devastation in 1906. Each catastrophe was followed by prompt rebuilding, with each rebuilding producing a new city better than the last. Earthquake, fire, and even plague could not stop its growth from 1847 to 1906.

Before the earthquake, San Francisco was by far the largest city west of Chicago and the ninth-largest city in the country, with a population of around 440,000. It was the largest port on the west coast, funneling the Pacific trade into the country and out. Large ethnic communities gave the city an international flavor. Many different newspapers and magazines were produced locally in many languages.

The best of everything was gathered together into one small corner of the world called San Francisco, which has four seasons, all of them spring. The early Chinese called it Gold Mountain, which was literally true, as gold was discovered on two separate occasions within the city itself.

Despite all of these qualities, the overwhelming lure of San Francisco since the Gold Rush has been San Franciscans. In 1908, historian Frank Morton Todd wrote, "Probably of all modern city communities, the San Franciscans are, as a class, the most careless, gay and free-spirited. Their most cherished right is the right to do as they please. They are known to be capable of hanging together in cliques, factions and parties, but their capacity for general concerted action had never been tested before 1906."

The original San Francisco City Hall was completely destroyed by the earthquake. Construction on the new city hall began in 1913 and was completed in 1915. The existing building was designed by Arthur Brown Jr.

Union Square is seen here in 1905, looking west from Stockton Street. In 1850, San Francisco mayor John White Geary presented 2.6 acres to the city. He left San Francisco in 1852, and the city council passed a resolution making it public property for all time. The tall building to the left is the St. Francis Hotel at Powell, Geary, and Post Streets. The statue, *Victory*, by Robert Ingersoll Aitken, is in the center of the square.

Rincon Hill is seen here in 1905 looking north to the Selby Smelting and Lead Company, on the southeast corner of First and Howard Streets. Some of the ore from the famous mines in California and Nevada was treated there.

The domes of the California Hotel on Bush Street near Kearny Street dominate this 1905 photograph. A chimney from the hotel fell on the neighboring fire station and fatally injured Fire Chief Dennis Sullivan. Goldberg, Brown and Company's warehouse is on the left. The Hopkins Institute of Art and the new Fairmont Hotel are in the distance.

Standing tall among the waterfront buildings in the distance in these two photographs is the Ferry Building, designed by architect Arthur Page Brown, who modeled it after Garalda Tower at the Cathedral of Seville.

The Call Building is seen here one day before the earthquake. Market Street is on the left and Third Street is on the other side of the Call Building.

City Hall the day before the fire

This photograph shows city hall on April 17, 1906, the day before the earthquake.

O'Farrell St before the fire

O'Farrell Street is seen here right before the earthquake. The Call Building is in the distance. As the tallest building in San Francisco, at 18 stories, the Call Building is a good marker; look for it in subsequent photographs to figure out approximate locations.

The Palace & Down Market.

This pre-earthquake and fire photograph looks east on Market Street to the Ferry Building. The world-famous Palace Hotel is on the right. This is the present location of the rebuilt Palace Hotel,

at Market and New Montgomery Streets.

This photograph looks from Portsmouth Square at the Hall of Justice, on Kearny Street. Washington Street is on the left of the building. This building is another good marker to help navigate the ruins in later photographs.

Two

EARTHQUAKE

The earthquake struck at 5:12:06 a.m., and fires immediately broke out. The shock was felt from Coos Bay, Oregon, to Los Angeles, and as far east as central Nevada, a total area of about 375,000 square miles, approximately half of which was in the Pacific Ocean. The region of destruction extended from the southern part of Fresno County to Eureka, about 400 miles away, and for a distance of 25 to 30 miles on either side of the fault zone. The distribution of intensity within the region of destruction was uneven.

Of course, all structures standing on or crossing the rift were destroyed or badly damaged. Many trees standing near the fault were either uprooted or broken off. Perhaps the most marked destruction of trees was near Loma Prieta in Santa Cruz County, where, according to Dr. John C. Branner of Stanford University, "The forest looked as though a swath had been cut through it two hundred feet in width." In just under a mile, Dr. Branner counted 345 earthquake cracks running in all directions.

The earthquake was so strong that sensitive seismographs around the bay were either knocked from their supports or the records went off the scale, so they gave no information as to the actual earthquake movements.

Destructive effects were greatest in the immediate neighborhood of the fault zone, but there were places many miles from the San Andreas Fault where the earthquake destruction was greater than in places nearer the fault. Intensified effects were found in the alluvial valley region, extending from San Jose to Healdsburg. Santa Rosa, 20 miles from the San Andreas Fault, sustained more damage, in proportion to its size, than any other city in the state.

This photograph was supporting evidence for someone's insurance company. It gives credence as to what occurred in this room on the morning of April 18, 1906. Many letters tell of the dust and the great sound. There is even one report of a person watching a power plant blow up through their window who could not hear the explosion because of all the noise in their house.

This wood-frame home had buckled floors and plaster stripped from the lath after the earthquake. Most people were in bed at the time of the earthquake, and a few even slept through it. Houses stood or fell depending on where they stood and what type of ground they were on.

This wood-frame building at Ninth and Brannan Streets was torn apart when the filled ground under it liquefied. In studies of the day, more people died in wood-frame buildings than in brick buildings.

The Valencia Hotel, at 718 Valencia Street in the Mission District, was a four-story building. The bottom part of the hotel collapsed forward, and the upper part of the building moved out into the street. The policeman who directed the rescue work reported that all of the people on the first and second floors were killed. Only eight people escaped the collapsed hotel.

Rescuers pulled rubble away board by board at the Wilcox House, at 109 Jessie Street, to save the trapped victims. The rescuers continued their efforts until they were driven away by fire.

This man, wearing a derby hat and a black suit, pulls his trunk full of his belongings with a rope. He had likely packed his trunk in a hurry, and he is on the move because he probably lost his home and possessions in the quake. His trunk must be heavy, for he is bending low in order to move it.

The house to the right has a sign nailed to its front appealing for lime, cement, and lumber. The owner is ready to begin repairing his home.

The buildings most subject to destruction in the 1906 earthquake were the wooden structures built on alluvial soil or landfill in the city's eastern half. The majority of those killed in the earthquake were in wooden, not brick, structures. The houses seen here later served as fuel for the conflagration that arose in the aftermath of the tremor.

Along Embarcadero Street, the crack of the earthquake is prominently seen. It is a little-known fact that bay water was actually thrown over the street as a result of the quake. The water progressed more than a block inland.

The facades of many buildings, especially wooden structures with brick face work, collapsed in the earthquake. Well-built brick buildings actually did very well in the earthquake, and many remain standing today. Even poorly built brick buildings usually just threw off part of a wall, which was dangerous to pedestrians but otherwise left the structure still standing and its inhabitants still alive.

Sidewalks and curbs were damaged by the liquefaction caused by the earthquake. There is an advertisement on the right for Max Wilhelmy & Co., a carpenter and builder on Eighteenth Street. This street damage was common in the flat areas of the Mission District. When buildings were rebuilt later, a survey found that the ground flowed like mud in that area, and property lines had to be redrawn.

Digging holes to bury belongings to recover later was a common practice. Many letters tell of putting silver and other property in yards to be dug up later.

The side of the Belvedere Music Hall was adjacent to this large business building with visible damage to its front. This was a common type of damage, as walls often peeled off and fell to the ground. It has become common knowledge not to leave a building during an earthquake, to avoid getting hit by falling rubble.

The earthquake shifted the tracks of the Presidio and Ferries cable line on Union Street near Steiner Street. This was fill land over a gully that had slipped out toward the bay. This photograph looks west.

Three

FIRE

San Francisco was destroyed by fire multiple times between 1849 and 1906. Initially, volunteers made up fire companies that responded to the fires. But, as the frontier town became a city, dependence on volunteers was no longer satisfactory. In 1866, the volunteer fire companies were abandoned in favor of a paid, professional fire department.

Despite this, from its founding, San Francisco was built to burn, and the fire department's chief, Dennis Sullivan, entertained no illusions about so-called fireproof buildings and the wide streets that were supposed to act as firebreaks. He was also aware of the poor water supply with low pressure in the fire hydrants, and the abandoned and deteriorated underground cistern system—a relic of the volunteer fire department days.

Spring Valley Water Company's system of mains for fire hydrants was considered barely adequate, and the 57 underground cisterns, left over from the Gold Rush days, were no longer considered by fire insurance companies to be a factor in fire protection. Many cisterns had been filled with trash and garbage, or utility companies had run pipes or conduits through them, although most still contained some water. On top of that, the locations of the cisterns were only known by the oldest San Francisco firefighters.

In 1906, firefighters were on duty 24 hours per day, in 10-day cycles with three days off per month. Firefighters could eat at home if it was near their assigned station, but they were required to sleep in the firehouse. However, most were not married and took up permanent residence in the firehouses. The uniformed strength of the San Francisco Fire Department on April 10, 1906—the most accurate figure that can be found—was 575 men.

The amount of fire department manpower on duty on April 18, as well as the adequacy or inadequacy of the water system, is somewhat academic, because the department simply never had a chance. Too many disasters happened simultaneously: at exactly the same time that thousands of people were caught in wreckage, damage to fire stations trapped the engines and caused the horses to run away, and the telephone and fire alarm systems were destroyed. All of these factors would have exhausted, disorganized, and fragmented the resources of the San Francisco Fire Department, even if there had been no major fires.

Each station of the San Francisco Fire Department acted on its own, as communications from headquarters was out. From some stations, the horses fled and the men pulled out the steamers and got to work. Some reported that they had to make hard decisions about whether to rescue people trapped in collapsed buildings or fight the fire.

Fires were starting all over the city. At the time, with wood and coal stoves in most buildings, any building that fell immediately started a fire. The most fires occurred on fill or made ground, where there were more buildings down, but fires occurred in all sections of the city. This image shows the fire on Third Street heading toward Market Street.

Since the cable cars were not in service, many citizens walked on them to view the progress of the fire. Most people watched the fire not knowing where it would stop.

The dome of the city hall is seen here in the center, with fires from south of Market Street merging and progressing north into the downtown area. The fire on the far left was known as the Ham and Eggs fire, as it started from cooking inside a home.

The spreading fire is seen here heading north toward the North Beach area. There are maps showing the directions of various fires, which were at the mercy of the wind. As fires got larger, they developed their own wind. Many people wrote about the sound of air being sucked into the fires. Readings of the wind swirling around the fires from military ships near the shore showed gale force.

Crowds walk through the smoke towards the Ferry Building to get away from the city.

This south-facing photograph shows the fire crossing Market Street at Grant Avenue. Troops watch helplessly as Chinatown burns and the flames continue to advance, building by building. This same fire climbed Nob Hill that evening and destroyed the mansions of San Francisco's wealthiest citizens.

People watch the fire move from block to block. In letters, some people wrote about going home for meals and then coming back to the streets to watch the fire until it reached their own houses.

This photograph looks southeast from the Mark Hopkins Home on California Street at Mason Street, with the Call Building in the center. The tallest building west of Chicago at the time, the Call Building is still in use today. Note the chimneys down on the buildings on the bottom right. It was estimated that 90 percent of the chimneys in the city were damaged in the earthquake.

This fire is being fought with water from a fire department steamer. The hose lines are charged and a stream of water is seen in the distance. The stories of having no water to fight the fire are generally wrong. The battle to save what was left of San Francisco lasted seven hours. Delirious firefighters collapsed in the street, only to be dragged from danger by refugees. Others rolled in the gutters to keep their melted rubber burnout clothes from sticking to their bodies.

The Ferry Building is seen here, looking west from the bay. The building was known as the Statue of Liberty of the West, because before the completion of the bridges across San Francisco Bay, it was the gateway to the city. It was erected on a foundation of piles, and at its July 1898 opening, it was hailed as the most solidly constructed building in California.

In this southeast-facing photograph, firemen in the distance throw water on burning homes on Market Street, a few blocks from Castro Street. Note the large crowd cheering them on. The fire was stopped in this area by citizens tearing down homes so the fire could not spread.

As fires began to run together, people saw a need to leave. At first, they just left their local area, but the whole city seemed to be on fire. The Ferry Building and the ferries, where everyone attempted to go, were crowded for many days.

Four

AFTERMATH

As soon as the flames died down, the great challenges of maintaining order, feeding the hungry, and sheltering the homeless began for the civil and military authorities. Although more than half the population had abandoned the city, thousands were left, stranded in their homes and in parks, military reservations, and the desolate western sand dunes.

San Francisco mayor Eugene Schmitz gave the approval to continue Gen. Frederick Funston's mobilization of the military in order to maintain order. Strict measures were enforced to guard against looting and to prevent outbreaks of pestilence. There was a need to establish camps, institute sanitary measures, and develop an orderly system of food distribution.

San Francisco's population was approximately 440,000 at the time of the earthquake. Southern Pacific's railroad evacuation alone accounted for the movement out of the city of more than half the population. Given that an additional 20,000 to 30,000 were evacuated by the Navy from the area of Fort Mason, it may have been one of the largest evacuations in history. It should be noted that these figures do not account for the passengers who fled the city by ferry. Oakland received the majority of people and cared for all it could, with those who could be forwarded to other places sent away on trains.

Between 6:00 a.m. on Wednesday, April 18, and Sunday night, the Southern Pacific ran 129 trains, with more than 900 cars, to the main line and local and eastern points, carrying refugees from San Francisco free of charge. During the same period, a total of 739 trains, with 5,783 cars, were run from Oakland. The number of people carried exceeded 225,000.

After the devastation of the earthquake and fire, San Franciscans faced the daunting task of resuming their daily routine. Living in tents and later in small shacks, residents cooked in the streets while children attended school and business people reestablished their trades. In spite of the ruins surrounding them, everyone was determined to carry on as before.

A crowd gathers on the streets below Nob Hill to see the results of the fire in the area. Looking west, this view shows the Fairmont Hotel in the center at the crest of the hill.

This general street scene shows the devastation of the fire. When it was safe to return, visitors came to the city to walk among the ruins.

This photograph shows the beginning of the cleanup process. Many people told of walking down the street and being told by police or soldiers to do some cleanup work. Some unlucky ones did this kind of work many times, going from one section of the city to another.

This view of Market Street looks east toward the Ferry Building, on the far right in the distance. People look around at the devastated landscape and watch the food and trinket vendors at work. Many people bought mementos of the fire and earthquake.

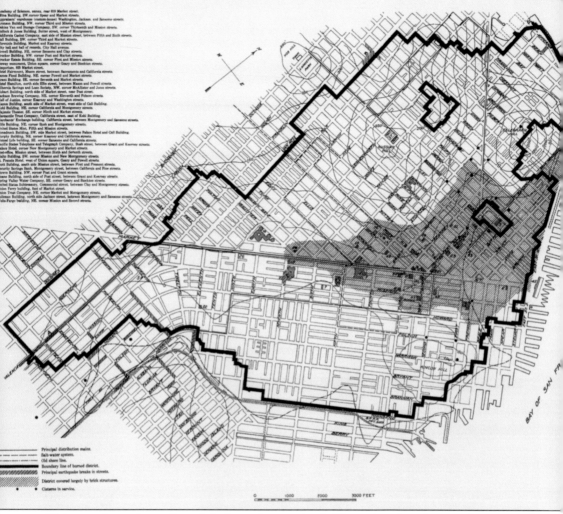

This map shows the outline of the burned zone, clearly illustrating that the waterfront was safe. The Navy saved this area, as the fire department was never there. Luckily, this allowed the docks to be in service for relief efforts.

Five

COOKING IN THE STREETS

The full magnitude of a disaster is never known until the danger has passed and the survivors must attend to the tasks of daily life. After the earthquake, San Francisco residents quickly tired of eating emergency rations and longed for proper meals, but everything was in confusion. Even if a house had escaped destruction, its chimney was probably lying in ruins. Only a meager selection of food was available for purchase, dishes were shattered, kettles had been commandeered to carry water to battle the fire, and high winds often extinguished the cooking fires of those still condemned to cook in the streets.

However, for ladies of social standing, the cruelest indignity was the fact that their cooks and domestic help had left to assist their own families. Many Nob Hill matrons were forced to learn the art of cookery over an open flame with a couple of salvaged saucepans and a set of silver spoons. Katharine Hooker, whose family home at Washington and Powell Streets had burned to the ground, described her fellow sufferers: "On stoves, from a kitchen range to a laundry flatiron-heater, or upon small furnaces improvised from bricks and bits of old iron, the cooking went on, and the cooks were not less diverting than the kitchens. Inexperienced householders tried sorry experiments, and the gentleman whose only suit of clothes combined a frock coat and a high hat assisted the millionaire whose servants had absconded, as they bent together over the little messes they stirred."

People moved their woodstoves out to the curbs in front of their houses. These outdoor kitchens became the social centers, and gossip reigned supreme. Everybody knew what the other people were cooking, and they could exchange formulae for what they were going to feed their old man that evening. However, there were certain hazards to this situation, namely the wind and the dust. The dust was not just ordinary dust, as this was a city with horses, so everybody had their meals with similar flavoring—not to mention possible unwanted extra nourishment.

Cooking in the street became commonplace throughout the city. Woodstoves were removed from the house and used in the streets for safety. Stove and fireplace use was not allowed in structures until they were inspected.

The Hoffman Café, at the southwest corner of Market and Second Streets, was the inspiration for this makeshift street kitchen.

This man, identified as a Chinese houseboy in the employ of Sarah D. Hamlin, the principal of the Hamlin School, prepares a meal in the street. These miniature stoves were made from five-gallon oilcans.

This woman has a roof over her head and a place to cook in a makeshift shelter dug out of mud and manure in Mission Dolores Park, near Eighteenth Street. The park is bounded by Eighteenth, Dolores, Church, and Twentieth Streets.

The following is excerpted from an April 23 letter written by Ernest H. Adams, a sales representative of the famed silverware manufactures Reed and Barton: "The city is under martial law and we are living on the government, or at least many are. As soon as the goods were safe, I cleaned out the nearest grocery store of canned goods and we are living in a tent, cooking meals on a few bricks piled up Dutch-oven style. Will endeavor to get into [the] city tomorrow, but every man caught in town is placed at work clearing the streets and they are kept at work until they drop."

This food must smell good, as a crowd has gathered. Note that these well-dressed people did not bring any chairs, as they would generally take their meals from the street into the house to eat.

This table with a cloth and plenty of chairs must have been very inviting. This camp was in one of many parks that were used to house the homeless. To clear their houses or tents of flies, people put a tablespoon of cayenne pepper in a pan over the fire and let it burn, opening doors and windows until the flies all disappeared.

This "Restrant" in the ruins was open for business. With so many sightseers around, many residents opened small businesses to serve the visitors. Most had basic fare like coffee and simple breads. It being 1906, there was of course no electricity or ice.

Hot coffee, sandwiches, and doughnuts were available in this tent in the ruins. These tent cafés provided a sense of community, as well as a brief respite from the continuous cleanup and rebuilding efforts.

Food distributed in the street was common in all sections of the city. Provisions were hauled by wagons of all sorts and distributed under the direction of military and civil authorities to any person who asked for food.

This bread line had a long wait. The clothes and hats on those waiting look like their Sunday best. For several days, all classes of people were forced to seek food supplies from contributions donated by various places outside of the city.

With the massive recovery effort underway, workers were sorely needed. Work registration posts were erected around the city. Signs like this one, reading, "All men and women able to work must register at once," were a common sight in parks, squares, and even amongst the rubble. This photograph was taken in Golden Gate Park, where there were some large camps.

In this typical food line, people wait for rations of supplies from a tent. This station provided food and clothing as needed for 1,200 people, including many refugees encamped in parks, other refugees sheltered in houses, and several permanent residents who lived within the limits of the station.

Sacks of flower and meat are prepared for distribution. The city was divided into supply districts, or sections, with each section run by a civilian chairman and an Army officer. Each section was numbered and further subdivided into food stations, which were also numbered.

A large number of workers eat at a community kitchen. There were many such kitchens in the city to take care of those who worked, which were set up similarly to company kitchens in military camps.

The homeless reaching for eggs.

These sailors distribute eggs to a crowd of citizens. Because eggs could be sent by train, many towns had egg drives for San Francisco. Passengers on fast trains saw flying freights, with every car labeled "relief," pass by them, while the passenger trains took the sidetrack.

This wagon handed out clothing to crowds of people in the Marina. Mount Tamalpais is on the right in the background, and Fort Mason is on the right. The gasholder in the center is near Bay and Laguna Streets.

Soldiers give out food in Golden Gate Park. Under the supervision of the National Red Cross and the military forces in San Francisco, the distribution of food and clothing was gradually systemized.

These boxes of food were part of a distribution station run by the military on Jefferson Square. The soldiers were armed, and did enforce good manners on the public.

Graveyards were used as camps, as can be seen here. Even the military graveyard in the Presidio was used for the living. Some of the earthquake's victims were also buried there.

Ongoing food distribution is seen here on Fourteenth Street near Market Street. This military station gave out food and kitchen utensils.

Bread lines run by the Signal Corps are seen here. Notice the automobiles, which were taken over by the military and used instead of horses. A number of letters from all military units gave glowing accounts of the automobiles. In some cases, the owners of the automobiles were paid to drive and service the machine. One such driver got paid $100 a day.

Many church and private groups also provided relief supplies. Helpers are seen here coming out of a station with cans and bagged bread.

Sacked potatoes and other food stores arrived in large quantities by boat at Presidio Wharf. They were contributed promptly by cities and towns near San Francisco Bay, and then hauled by wagons and distributed under the direction of military and civil authorities.

This makeshift soup kitchen was set up in the street to feed the hungry. The work of distributing supplies and maintaining free kitchens was done by volunteers, and a house-to-house canvas of the station territory was performed by other volunteers in order to supply information to the Red Cross, which could then try to find work for those desiring it.

Six

MILITARY

Gen. Frederick Funston had 1,500 troops in the city by 12:00 p.m. on the day of the earthquake, an armed force that more than equaled the combined strength of the police and fire departments.

"I have no doubt," General Funston wrote, "and have heard the same opinion expressed by scores of citizens, that had it not been for the prompt arrival of the large force of Regular Troops, who were acting under orders to shoot all looters, the saloons would have been broken into, and then the crowd, becoming turbulent, would have begun sacking the banks and jewelry stores. The city police, however brave and efficient, would have been totally unable from mere lack of numbers to have dealt with such a situation."

The panicked people who milled in the streets seemed to accept the troops as an expected precaution. There was a general assumption that martial law was in force, even though it was not. At best, San Francisco was an occupied city under military rule because the Army was the only disciplined, cohesive force operating within the city. In this case, the erroneous assumption on the part of the people that some form of military rule was in force was as effective as any declaration.

In those early hours of the disaster, Mayor Eugene Schmitz's and General Funston's perceptions of danger quickly changed. The mayor's order to police chief Jeremiah Dinan to close the city's saloons was a stiff one for the overworked, uncoordinated, and disorganized 700-man force. General Funston telegraphed the War Department for enough tents and rations for 20,000 people; within an hour, he raised that figure to 100,000.

The regular policemen were the steadfast hope of the people, and they policed like heroes and ruled like fathers. In the central part of the city, they checked the outrageous famine prices offered to starving people. When Thursday morning broke, lines formed before the stores whose supplies had not been commandeered.

This armed military group patrols in the center of the burned-out city. On the last day of the fire, people noticed an increased number of soldiers patrolling the streets. They later learned that the extra troops had been brought from Army posts in Montana, Utah, and Wyoming.

With San Francisco in desperate need of security, food, medical aid, and supplies, the military was deployed throughout the city. It was calculated that it would take one soldier for every 50 people in the city to properly provide aid.

In this camp, police and fire departments worked together to provide meals for hungry citizens. The earthquake had a numbing effect on both departments, neither of which was ready for such a catastrophe. Once all communication was lost after the quake, the men were forced to rely on each other and those around them.

This provision and supply station in Golden Gate Park distributed stoves and food to refugees. Two companies of the 22nd Infantry were assigned as a working police force and put in camp near the park lodge. They were very useful in the early days of the emergency, when there was a good deal of confusion. After a few days, the soldiers were withdrawn, replaced by two companies of the 14th Infantry. Later, Troop L, 14th Cavalry, was assigned to the park.

The sign outside this military distribution area in Golden Gate Parks reads, "Positively none but employees allowed in here."

The military guards a line of refugees three blocks long, awaiting the distribution of meat and supplies on Sanchez Street, east of Fourteenth Street in an unburned part of the city.

In addition to the regular military troops, state militia were mobilized by the governor. They were far less disciplined than the regular military, and were involved in many incidents of misbehavior, acts of drunkenness, and even the wanton shooting of innocent people. Unfortunately, in their criticism of the military, the populace had trouble differentiating one uniformed figure from another.

By 8:00 a.m. on the day of the disaster, less than three hours after the earthquake, 400 officers and enlisted men of the San Francisco National Guard had reported to the various armories in the city for duty, knowing immediately that such disasters called for their service.

The California National Guard Command Commissary hands out supplies to a line of refugees on Duboce Avenue near Church Street.

The California National Guard prepares for lineup. This camp provided police service and relief supplies for the area around Church and Market Streets. The national guardsmen simultaneously acted as soldiers, firemen, doctors, nurses, and purveyors of food to the hungry.

Despite the chaos following the quake and the fire, the citizens of San Francisco carried themselves with dignity and grace, as noted later by Gen. Frederick Funston in *Cosmopolitan* magazine. "Through all this terrible disaster," he said, "the conduct of the people had been admirable. There was very little panic and no serious disorder."

The California National Guard Command lines up near Church and Market Streets with their full band.

The 1st Infantry, California National Guard, is seen here at the upper Market Street camp, near Church Street. Some of the officers in service in San Francisco during the disaster went on to become generals in World War I.

The California National Guard Command is lined up for this group photograph. They acted as police and additional fire protection for the area. According to reports in many public and military letters, the San Francisco Police were ineffective for weeks, and some sections of the city never saw a policeman for up to two weeks.

Gen. Henry G. Mathewson (left) recruits help for the 1st Infantry of the California National Guard after the disaster.

The Regimental Commissary of the California National Guard is seen here. The entire California National Guard was ordered to duty after the earthquake. The 7th Infantry Regiment (now 160th Infantry) was sent from Los Angeles to Oakland, the 5th Infantry Regiment (now 159th Infantry) was mobilized in Oakland, and several companies of the 5th Infantry Regiment were sent to San Francisco.

The Young Men's Christian Association (YMCA) set up this tent for the 1st Infantry, California National Guard, at the upper Market Street camp, near Church Street.

The California National Guard's upper Market Street camp is seen here. To the left is the present location of the San Francisco Mint. Market Street is out of the photograph to the right, and Church Street is behind the camera.

The 1st Infantry, California National Guard, is seen here on parade, probably in the upper Market Street area, near their camp.

The Quartermasters Department of the 1st Infantry, California National Guard, is seen here on Duboce Avenue, looking west. The building in the background is on the northwestern corner of Church and Duboce Streets and is still there today.

The military was tasked with providing a sense of order in the rubble. According to a telegram from Maj. Gen. Adolphus Greely to Secretary of War William Howard Taft on April 26, despite "fearful conditions indicating demoralization dishonesty and deception," Major General Greely was confident that these conditions "[did] not yet acknowledge incapacity to bring order and efficiency out of existing chaos."

Before the disaster, automobiles were generally seen as toys for the rich. After the quake, however, cars were instrumental in putting out the fires that continued to blaze for days. As noted by Howard T. Livingston, "The fire-fighting had demanded the use of all available vehicles—horse-drawn and mechanical—and most of the automobiles in San Francisco, including new ones still in dealers' showrooms, had been pressed into service." In the recovery efforts, cars quickly proved their superiority to horses.

Capt. Leonard Wildman relayed the first orders to the troops on Wednesday morning by driving the single automobile that was in the commission of the city government at the time of the fire. Driving to the dock at Folsom Street through the burning district, he caught the steamer *McDowell* and sent orders for the troops to come in from Alcatraz and Angel Island.

A boy shows off his two-dog-powered wagon on Church Street between Hermann Street and Duboce Avenue.

The soldiers were such a welcomed sight to the survivors that they were requested to remain past the withdrawal date of July 1, 1096. In a telegram, Secretary of War Taft stated, "I have conferred with General Greely and he states that it is agreeable to him to comply with the wishes of the citizens of San-Francisco provided the Secretary of War has no objections. I respectfully request that the Army be continued in its present usefulness for at least sixty days beyond July first."

Policemen camp at Portsmouth Plaza, receiving food and even a shave (center). The Hall of Justice was on the east side of this square. The hall housed the jail and was damaged by the earthquake. Prisoners were removed after the cells were opened just ahead of the fire.

Seven

REBUILDING

With the smoke clouds overhead and the ashes but half-cooled beneath its feet, the indestructible spirit of San Francisco asserted itself. In fact, some buildings were laid down over the ruins only to catch fire from the smoldering remains. The owners of businesses, watching them burn, planned greater buildings by the light of the fire. It was the optimism of faith and courage, not bravado and carelessness.

Systematically and immediately, the quickly organized Citizens' Committee, city, federal, and state authorities, railways, and other agencies of power began the work of relief and reconstruction. Peace and order were maintained by the military and later by the San Francisco Police Department. The main streets were soon cleared for traffic. The businessmen laid aside personal affairs in the hour when they were needed most, and instead gave their time and thought to the common good. The railroads, unsolicited, gave free transportation to the fleeing refugees and carried the contributions of the nation for the relief of San Francisco. The president of the Southern Pacific and Union Pacific, Edward F. Harriman, hurried westward as fast as steam could carry him to take part personally in the work of relief and restoration. He wired ahead to turn all the lines over without charge to the work of relieving San Francisco.

In an April 18, 1908, article for the *San Francisco Examiner*, H. Morse Stephens, a professor at the University of California at Berkeley, wrote:

San Francisco is imperishable; she is not made of bricks and mortar or reinforced concrete; she is a temperament, a sunny, vivacious temperament, which with gay courage faces great disaster and overcomes it and which rebuilds with courage of heart what she saw burn up with gayety of heart. Her temperament cannot be analyzed, it can only be felt; we can admire her for what she is doing in the present to build up again her homes, her hotels, her office buildings and her stores, but we love her for the way in which she met with undaunted courage and with a bright smile the disaster of two years ago, which would have crushed into eternal depression any other city upon the face of the globe.

This was the cover of the special edition of the *San Francisco Examiner* six months after the disaster, on October 21, 1906.

These cable cars were used to house some of the people displaced by the disaster.

This view looks west at running streetcars and new tracks being laid to remove the rubble. The Call Building is on the left, and the Twin Peaks are in the distance. The Call Building was gutted by the fire, but was renovated and is still in use today.

These earthquake shacks were produced in large numbers and placed in parks or open areas. They were sold after they were no longer needed and then moved out of public areas, becoming homes all over the city, with some combined into multi-room buildings.

W.S. Snook & Son Plumbers and the Joe Valvo Barber Shop awaited customers soon after the earthquake and fire.

This Western Union telegraph and cable office opened for business in a shack after the earthquake. For the first time in its history, San Francisco was absolutely without telegraphic communications with the outside world for three hours on the morning of April 18. Ceaselessly, through every hour of the day and night, messages were speeding to all the most remote sections of the globe, so to have the busy workings of the "tickers" stopped for so long was an event in telegraphic history.

These cable cars were set up as homes along California Street between Third and Fifth Streets in the Richmond District. The wheels were removed and extensions were made to increase living space.

Waltz Safe & Lock Co. is seen here after it reopened with a temporary outdoor office. The business could not be reopened for many days in the immediate aftermath of the fire, because safes had to cool down before they could be opened, as the contents could burst into flame if opened when the steel was still too hot.

The Fresno Relief and Information Bureau set up after the disaster to persuade people to move to Fresno. It is estimated that roughly 220,000 people—about half of the city's population—fled the city and did not return. This mass exodus helped populate other areas of the state.

School was held in tents in various parts of the city, as most of the schools were destroyed in the earthquake and fire. The first cash contributions for schools came from the boys and girls of Broken Arrow in Creek Nation, Indian Territory, a beautiful illustration of the kinship of the peoples of our beloved country.

The fire stopped on the west side of Van Ness Avenue; the east side of the street went up in smoke. Stores opened quickly in the area, creating a shopping area until the downtown was rebuilt. This image looks north, with the western additions on the left.

Union Square is seen here with rebuilding underway. The St. Francis Hotel was behind the camera on Powell Street, and the photograph looks southeast, with Geary Street on the right and Stockton Street running from left to right.

Streetcar workers take a break in the Outer Mission. Note that the ground has moved the level of the buildings. In some places, the ground sank up to six feet. Some homes were re-leveled and continued to be lived in.

San Francisco in Ruins.
April 18. 1906.

1377

EDDY & MARKET

Running the first Street Car, after the calamity.

This postcard shows the running of the first streetcar after the earthquake and fire.

Many people dragged trunks and bags behind them, often with pets of all descriptions, including goldfish and birds. After the disaster, it was noticed that many exotic birds were free, and it was assumed that many people just let them out instead of taking them when they fled their homes.

These people are heading to the Ferry Building to leave and go anywhere. People leaving the city were given free passage anywhere within the state. California's counties got considerable boosts in population, and most people who left never came back to the city. San Francisco's population before the earthquake was around 440,000, and it dropped within days to 220,000 because of people fleeing the city.

Earthquake fissures and streets sunk five feet, as seen here, were common sights south of Market Street. This photograph shows Eighteenth Street between Howard and Folsom Streets.

The burned-out ruins of Lincoln School, at Fifth and Howard Streets, are seen in this north-facing photograph. The Call Building is in the distance on the right.

Block after block in all directions looked the same. Sightseers came by ferry from Oakland to walk the streets. Many tent restaurants catered to the visitors and to the people working to remove the rubble.

Earthquake S.F. Apr 18th. Tent City Dolores & 20th Str. N.W. on edge of Fire

This tent city was set up at Dolores and Twentieth Streets. This photograph looks northwest from the edge of the fire zone in the Mission District. Today, Dolores Park covers the area on the right side of the photograph.

Copyrighted 1906 by C.P. Magagnos. Alameda Cal.

This photograph shows the city on fire from the bay, with the Ferry Building near the center. The fire was generally out after three days, but the rubble smoldered for months. In fact, months later, some new construction built on top of the rubble even caught fire.

Camp Seven in Golden Gate Park is seen here set up with tents from the military. The military constructed these camps, supplying tents, tools, and necessary conveniences for cooking and sanitation, after the manner of military camps. This arrangement continued until May 7, when the camps were turned over to the city.

People wait in line for food being distributed from a tent. Soon after the disaster, stores of food began to arrive in large quantities by boat, contributed quickly by nearby cities and towns.

This east-facing view of Market Street from Fourth Street shows the streets still full of rubble and people. The Call Building is on the right.

These neighborhood women set up kitchens in the street when they had to remove the stoves from their homes. While they cooked outside, they still went inside their homes to eat, which is why there are no chairs in the area.

Temporary offices for the sheriff, the county clerk, and the district attorney were set up at California and Webster Streets.

The first business to open in San Francisco after the fire was this open-air barbershop on the Mission Street side of the US Mint, at Fifth and Mission Streets.

The fire-gutted Hibernia Bank at Jones and McAllister Streets housed the San Francisco Police Department's Harbor Station for a few months after the disaster. The bank building was reconstructed, and the exterior remains essentially the same today. The spire of St. Boniface Church on Golden Gate Avenue is in the background.

What must have been very hungry horses pull a wagon filled with refugees to the Ferry Building. Local entrepreneurs began private transit service between the refugee camps and the ferries because of the destruction of the streetcar and cable systems. This wagon charged 25¢, five times the normal fare.

Fire Engine No. 5 and Fire Truck No. 2 are at their temporary quarters on the waterfront after the fire. After the disaster, the Navy established a ship whistle system to alert the fire department of any waterfront blazes.

The collapse of the San Francisco Gas and Electric Company's gasworks during the earthquake sounded like an explosion to the people in North Beach.

Fillmore Street was never burned, and featured the last shopping area in the city. The temporary police station on the right was at the corner of Bush and Fillmore Streets.

The Emporium, which had been on Market Street, was reopened on Van Ness Avenue after the fire. When its former location was rebuilt, the store moved back.

Safes were hauled to the streets from buildings and left to cool. Two weeks were required before they could be opened again, any earlier and owners risked the instant immolation of their contents. The safes were pulled into groups and guarded. Most safes were a loss anyway, for many of them could not withstand the great heat they endured.

The earthquake snapped off the top of the smokestack at the cable car powerhouse at Washington and Mason Streets. The falling smokestack tore through the roof of the structure.

The tall smokestack of the United Railroads powerhouse at Oak and Broderick Streets collapsed, wrecking several cable cars inside the car barn.

A string of burned cable cars is seen here along Pacific Avenue at Polk Street. Cars that did not burn were used as temporary housing.

The Butchertown piers dropped into the bay near Hunter's Point.

This post-disaster view looks north, showing the destruction of Chinatown and Telegraph Hill. The Hall of Justice is on the right on Kearny Street.

San Francisco is seen here from a captive balloon one month and 11 days after the earthquake and fire. The aerial view below looks south and shows how the great majority of the city was left to ashes, with city hall on the right and Mission Bay on the left. Mission Bay was filled in by rubble.

The terminus of the Larkin Street Railway on Ninth Street is seen here, showing ground movement on the tracks and distorted buildings.

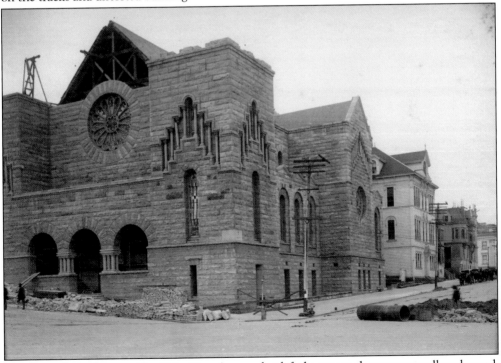

The fire took out the east side of Van Ness Avenue, but left the area to the west generally unburned. This photograph shows the southwest corner of Broadway and Van Ness Avenue. The church is still there today.

Children amuse themselves with a pet goat in the rubble on the west side of Telegraph Hill.

All the military tents available were set up in the Presidio, and the troops were turned out of the barracks to bivouac on the ground. In these shelter tents, they first placed the sick, then the more delicate of the women, and then the nursing mothers. The military provided food and medical support for all.

This simple camp was set up in a park. Note the stove and park benches being used around this camp. Men were making little tents of sheets and blankets, held up with sticks taken from collapsed buildings.

School was conducted in tents furnished by the military authorities in Golden Gate Park for all the schools that were destroyed. Soldiers served as truant officers; here, a drill sergeant puts the class through exercises.

This photograph looks from Russian Hill down over North Beach and on to the waterfront. Mount Tamalpias, in Marin County, is on the left, and Alcatraz is near the center.

This temporary home was likely set up on the side of Twin Peaks area.

Here, a carpenter visits with family and a friend in the temporary shacks set up after the earthquake and fire.

This tent home was set up among the unburned homes, probably on the north side of Telegraph Hill. The tent is from the military. Every tent in the country that the military had was sent to San Francisco.

This man set up a well-stocked general store in the rubble.

Swift Bros. Lumber, in the Mission District, is seen on the right in this west-facing photograph, with Twin Peaks in the center in the distance.

This pre-1906 photograph shows the California Baking Co. at 1501 Eddy Street, on the southwest corner of Eddy and Fillmore Streets.

Military guards were stationed at the doorway of the California Baking Co. on the day of the fire, where free distribution of bread took place at various times throughout the day. Men, women, and children stood in line for blocks, quietly waiting their turn, but at the bakery door only the vigilance and firm bearing of the sentry maintained order.

The California Baking Co. is seen here, improved and rebuilt 14 months after the fire. The company deserves special commendation for getting promptly to work to supply bread while rumors of approaching fire were rife. They began at once to repair the damage done by the earthquake and were an excellent example to the people who came to them for succor with or without payment, according to a Navy report. The building was severely damaged by the earthquake but continued operation throughout the relief period.

The front of the California Baking Co. retail store on Fillmore Street is seen here, with samples in the windows for all to see. This company was well respected in San Francisco for many years.

The interior of the California Baking Co. retail bakery is seen here, with the staff behind the main counter waiting to serve. The interior was lit by gaslights at the time.

The Original Creamerie Restaurant on Fillmore Street had a separate dinning room for ladies.

This interior photograph of the Original Creamerie Restaurant shows the tables in the dining room and the staff ready to serve.

Kitchen cooks at the Original Creamerie prepare meals. Note the servers waiting on the right.

Stands were set up all over the city, selling mementos and items found in the rubble. These visiting sightseers look at a stand on Market Street.

These were two popular comic postcards printed after the earthquake. Thousands of these were sent by refugees to anxious relatives on the east coast to let them know all was well.

Eight

ENTERTAINMENT

Postcards, books, magazines, and newspapers covered the lighter side of the disaster, while music, poetry, and theater gave much to consider.

After the earthquake and fire, San Francisco became exceptionally busy and more interesting. The task was to create a great city from a mass of unrecognizable ruins. That the new city would be a larger and better one than the old felt like a certainty.

While the people of San Francisco could not exactly consider themselves fortunate, they did suffer an experience very nearly unique in the history of the world. People rejoiced at the opportunity to take part in a great feat of human energy, a feat in which the pick and pinchbar were as wands wielded by so many magicians.

The woman seized a wrench lying at her feet and, leaning over, struck the detective a savage

The cover of the *New Nick Carter Weekly* comic book portrayed precisely the image of San Francisco that its business community did not want to project to the world. It was printed exactly four months after the earthquake, on August 18, 1906, in New York.

Famed artist Maynard Dixon was commissioned to paint the cover of the Southern Pacific Company's *Sunset* magazine, which would commemorate the birth of the New San Francisco.

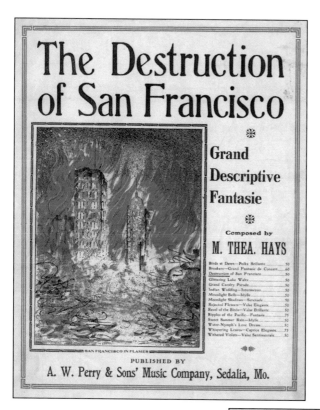

This sheet music for "The Destruction of San Francisco" was published by A.W. Perry and Son's Music Company of Sedalia, Missouri.

This sheet music for "In San Francisco the Fair Will Be Best" was published by the Southwest Music Publishing Company of Oakland.

"San Francisco Memorial Song" is seen here. The sheet music was published by Wm. E. Cornwell of Los Angeles.

The sheet music for "The Stricken City" was published by the American Advance Music Company of New York.

THE REFUGEES' COOK BOOK

COMPILED BY ONE OF THEM

1906

The Refugees' Cookbook, "compiled by one of them," offers a glimpse of the sorts of dishes that were popular in San Francisco's socially conscious dining rooms at the time of the disaster. But the recipes also reflect a longing for normality and comfort. These were not the meals refugees actually ate while living in the camps, these were the meals they had eaten before the disaster, and what they longed to have as they huddled around the campfire, enveloped in fog.

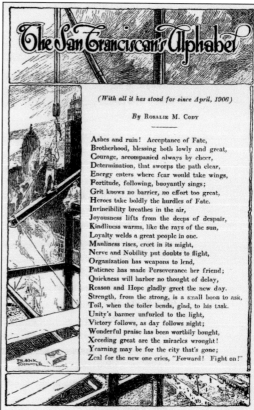

The San Franciscan's Alphabet

(*With all it has stood for since April, 1906*)

By ROSALIE M. CODY

Ashes and ruin! Acceptance of Fate,
Brotherhood, blessing both lowly and great,
Courage, accompanied always by cheer,
Determination, that sweeps the path clear,
Energy enters where fear would take wings,
Fortitude, following, buoyantly sings;
Grit knows no barrier, no effort too great,
Heroes take boldly the hurdles of Fate.
Invincibility breathes in the air,
Joyousness lifts from the deeps of despair,
Kindliness warms, like the rays of the sun,
Loyalty welds a great people in one.
Manliness rises, erect in its might,
Nerve and Nobility put doubts to flight,
Organization has weapons to lend,
Patience has made Perseverance her friend;
Quickness will harbor no thought of delay,
Reason and Hope gladly greet the new day.
Strength, from the strong, is a small boon to ask,
Toil, when the toiler bends, glad, to his task.
Unity's banner unfurled to the light,
Victory follows, as day follows night;
Wonderful praise has been worthily bought,
Xceeding great are the miracles wrought!
Yearning may be for the city that's gone;
Zeal for the new one cries, "Forward! Fight on!"

The San Franciscan's Alphabet was written by Rosalie M. Cody in June 1907.

Nine

INJURY, DEATH,
AND DISAPPEARANCE

On April 18, 1906, at 5:12:06 a.m., a large portion of the American west was shaken by a major earthquake, the epicenter of which was in Olema, in Marin County, California. This earthquake, on the San Andreas Fault, distorted the surface of the land by an agitation most leading seismologists later interpreted as being around 8.3 on the Richter scale.

San Francisco, the Queen City of the West, was in instant trouble, as were her neighbors to the north, south, and east. It took weeks for the dust from San Francisco's fallen buildings to settle, and three days and nights for the resulting fires to be extinguished.

After the disaster, perhaps because buildings were easier to account for than people or perhaps because they were more important to the statisticians, it was announced that 4.11 square miles of the city had been blackened and 28,000 buildings had been lost. Almost a half century elapsed before the official count of the dead quickly provided by the board of supervisors, 478, was challenged by members of the San Francisco Earthquake Research Project. The documentation will probably never be complete because the known death toll continues to grow with each year, as genealogists uncover letters from ancestors revealing the deaths and disappearances of loved ones. In addition, the transients or homeless whose remains were incinerated in the buildings that collapsed upon them will remain forever unknown.

A city street is seen here, with one of the nameless victims exposed for all to see.

The dead body of a fire victim is huddled in the ruins.

This burned body of a looter lay at the intersection of Post Street and Grant Avenue. According to police department records, the man had attempted to burglarize Shreve's Jewelers just before the troops arrived downtown. He was caught by the crowd and turned over to a soldier, who then shot him. His body was left to burn in the intersection.

The injured were moved from one place of perceived safety to another, as all of the hospitals downtown had to be evacuated. Some of the injured went to the parks and some went to homes, if they were lucky. Some were carried on chairs or even wooden boards.

One of many emergency hospitals in Golden Gate Park is seen here. These quickly set up hospitals treated thousands with the help of the military, which provided security and tents plus supplies from the Presidio.

Water is distributed at the Ferry Building. Water, which was only for drinking or cooking, was distributed by water wagons brought over on ferries from Oakland.

All of the city's hospitals were damaged, and only one did not burn. The injured were cared for in homes and in the parks. Here, a victim is delivered to a park for care.

A patient is worked on by a doctor in a tent city, most likely in Golden Gate Park. It soon became apparent that the regulation 108-bed field hospital the military would put up would be entirely inadequate for the number of applicants. A request was made for more, and a large number of quartermaster store tents and many hospital tents were supplied from the Presidio.

Victims removed from the Mechanics' Pavilion are unloaded at a temporary hospital in the Golden Gate Park panhandle.

Rescuers pulled rubble away board by board at the Wilcox House as they tried to save trapped victims. According to the *New York Times*, Fathers Hogan, Rogers, and Huber of St. Patrick's Church gave the trapped and dying victims last rites.

A San Francisco firefighter pulls a victim from the rubble of the Wilcox House.

HOSPITAL IN GOLDEN GATE PARK.   COPYRIGHT 1906, PILLSBURY PICTURE CO. NO. 39.

Another temporary hospital established in Golden Gate Park to treat the abundance of injured people is seen here. Doctors here reported that prostitutes from the Barbary Coast came to the hospitals to beg for morphine because their supply of the narcotic had been destroyed by the earthquake and fire.

A doctor and his staff of nurses are seen here in Alamo Square Park. The physical welfare of the cottagers was the camp nurses' special line of work. Each nurse or group of nurses had one cottage, which included a dispensary. In the cottages, the nurses assisted the camp doctor with dressings and dispensing drugs. They also slept in the cottages.

115

The exit doors along the Polk Street side of Mechanics' Pavilion were thrown open to begin the mass evacuation of the wounded before the fire arrived at the pavilion. Injured patients were taken on an agonizing ride to Golden Gate Park or to the US Marine Hospital on Lake Street. Many of the dead were left in the building to burn.

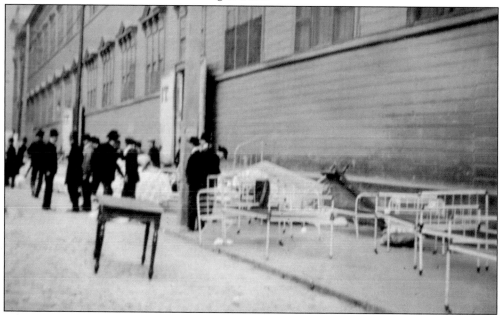

Abandoned hospital beds are seen here along the Hayes Street side of Mechanics' Pavilion after the evacuation. Only the patients deemed capable of saving were evacuated in time, and the rest were chloroformed or shot to prevent them from suffering the ordeal of being burned alive. According to Dr. Wolff, who was present for the evacuation, the medical staff went about their ghastly work in a state of shock, and even the policemen standing guard outside could not help but weep at the awful spectacle.

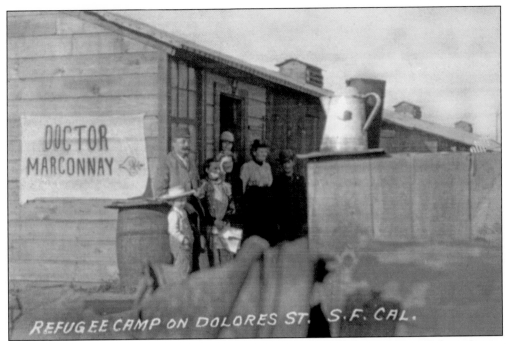

The temporary office of a Dr. Marconnay is seen here in a refugee camp on Dolores Street. Permanent refugee camps were under military control.

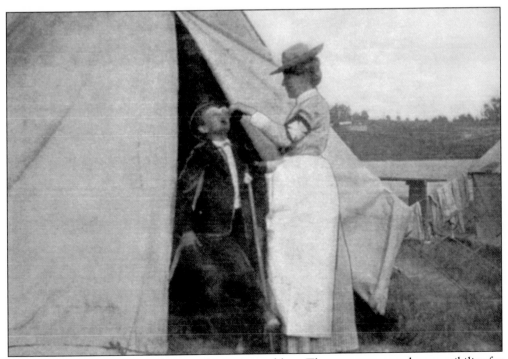

This relief camp nurse provides care for an injured boy. The nurses assumed responsibility for monitoring and isolating infectious disease cases after the Army left.

The Associated Charities relief station, seen here, provided services for all who could come. After the withdrawal of the majority of the camp nurses and the return to normal conditions, the regular Associated Charities nurses, with diminished help but no fewer patients, were busier than they were when the camps were in full swing.

Water wagons from Oakland provide water for a long line of thirsty citizens. Food and water were distributed at this location by the military.

Relief stations were set up in all areas of the city. These stations were vital because most of the stores could not get supplies, as the main distribution companies had been reduced to rubble.

Old men and "weaklings" were ordered by Gen. Frederick Funston to dig graves in Portsmouth Plaza, across from the Hall of Justice. The condition of dead bodies was becoming a major health problem and they required immediate burial, but the troops had no men to spare to dig graves, and the young and able-bodied men were mainly fighting on the fire line or were utterly exhausted. The old men did it willingly enough, but had they refused, the troops on guard would have forced them to.

This Red Cross tent was set up for the distribution of supplies in Jefferson Square.

Author Henry Anderson Lafler, seen here, wrote articles about the condition of the city for *Collier's* magazine on his portable typewriter, salvaged from his home at 612 Clay Street. He worked at this small table amid the fresh graves dug in Portsmouth Plaza. The ruins of the Hall of Justice are on the right.

More graves are dug as a soldier stands guard in Portsmouth Plaza. Many bodies disappeared in the collapsed buildings and were consumed by fire; others were buried in mass graves without any attempt at identification.

Coffins are lined up at Portsmouth Plaza, with the specter of the burned-out city in the background.

This photograph looks north from the southeast corner of Portsmouth Plaza, where graves are visible. It was ruled that every man physically capable of handling a spade or a pick should dig for an hour. When the first shallow graves were ready, the men, under the direction of the troops, lowered the bodies, several to a grave, and the strange burial process began. The women gathered about, crying. Many of them knelt while a Catholic priest read the burial service and pronounced the absolution.

Water was taken from anywhere it could be found, including from under cracked streets, as seen here. The military posted signs in a number of languages urging people to boil the water for health reasons.

Dead bodies are seen here awaiting burial in Portsmouth Square before the fire came through the North Beach area.

Mowry's Opera House, at Grove and Laguna Streets, served as the temporary city hall after the earthquake and fire. The old opera house was eventually destroyed by fire as well, but not until 1970.

Hamilton Park is seen here with a variety of tents, even circus tents, providing shelter. In the Presidio, one circus tent was used as a hospital.

These people made a camp under the cypress trees in Golden Gate Park. Every tent and field stove owned by the military anywhere in the country was sent to San Francisco. In one story about the tent city that sprung up in Golden Gate Park, the inhabitants needed toilet tissue so badly that they raided the library at Associated Colleges, near the park, and used paper from the books.

This photograph, taken near the Ferry Building, has been widely distributed in the more than 100 years since the earthquake and fire. Imagine what those eyes saw.

DISCOVER THOUSANDS OF LOCAL HISTORY BOOKS FEATURING MILLIONS OF VINTAGE IMAGES

Arcadia Publishing, the leading local history publisher in the United States, is committed to making history accessible and meaningful through publishing books that celebrate and preserve the heritage of America's people and places.

Find more books like this at
www.arcadiapublishing.com

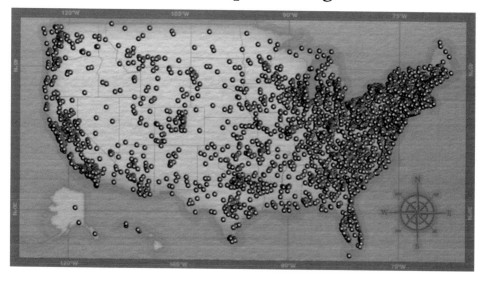

Search for your hometown history, your old stomping grounds, and even your favorite sports team.